Abdelhafid Mimouni

When Proton goes into Action

Abdelhafid Mimouni

When Proton goes into Action

ScienciaScripts

Cover image: www.ingimage.com

This book is a translation from the original published under ISBN 978-620-6-72803-0.

Publisher:
Sciencia Scripts
is a trademark of
Dodo Books Indian Ocean Ltd. and OmniScriptum S.R.L publishing group

120 High Road, East Finchley, London, N2 9ED, United Kingdom
Str. Armeneasca 28/1, office 1, Chisinau MD-2012, Republic of Moldova, Europe
Managing Directors: Ieva Konstantinova, Victoria Ursu
info@omniscriptum.com

Printed at: see last page
ISBN: 978-620-8-38550-7

When Proton goes into Action

Author : Dr. Abdelhafid Mimouni : Independent researcher in chemistry

Holder of a doctorate in chemistry from the University of Paris XII

(1997) and a Diplôme des Études Approfondies en Systèmes

Bioinorganiques

from the University of Paris XI (93), a Bachelor's degree and a Master's

degree in chemistry (91,

92).

Summary: This book takes an in-depth look at proton therapy, an advanced radiotherapy technique that uses protons to treat cancer. It begins with a presentation of the fundamental principles of proton therapy and the importance of the Bragg effect, which enables tumours to be precisely targeted while preserving healthy tissue. Through clinical case studies, we examine the promising results obtained for various types of cancer, highlighting the advantages of this approach over conventional radiotherapy. The book also discusses methods for calculating and modelling the Bragg distance, highlighting their impact on treatment optimisation. Technological innovations, such as miniaturised accelerators and the integration of advanced imaging, are discussed, as well as ongoing research into the Bragg effect. Finally, future prospects, including the integration of proton therapy with other treatment modalities, are presented, paving the way for precision medicine in the fight against cancer.

Plan :

Introduction

Proton therapy represents a significant advance in the field of radiotherapy, offering a targeted and effective approach to the treatment of cancerous tumours. This innovative technique uses protons, positively charged subatomic particles that have unique properties compared with conventional X-rays. Unlike conventional X-rays, which penetrate tissue with a gradual loss of energy, protons have a distinct energy distribution. This allows precise control of the dose delivered to tumour tissue, thereby reducing damage to surrounding healthy tissue. This specificity makes proton therapy a particularly promising treatment option for certain types of cancer, especially those located near sensitive organs such as the brain, spinal column or salivary glands.

One of the fundamental principles of proton therapy is the Bragg effect, which describes the ionisation peak that protons reach at a specific depth in the tissue. This phenomenon is crucial, as it enables clinicians to maximise dose deposition within the tumour while minimising exposure of healthy cells to radiation. A thorough understanding of the Bragg effect is therefore essential for optimising treatment protocols, adjusting dose parameters and improving clinical outcomes for patients.

The aim of this book is to provide a detailed exploration of the Bragg effect in the context of proton therapy. Not only will we examine the underlying physical principles and the calculations required to determine the Bragg distance, but also the clinical applications of this innovative technique. Drawing on case studies and recent research, this book aims to enlighten healthcare professionals, researchers and students about the importance of the Bragg effect and its potential in the evolution of cancer treatment. We hope that this resource will contribute to a better understanding of proton therapy and its benefits, encouraging its adoption and development in the contemporary medical landscape.

Chapter 1: Fundamentals of Proton Therapy

1.1 History and development of proton therapy

Proton therapy, a revolutionary therapeutic modality, was conceptualised in the 1940s, a pivotal period marked by major advances in the fields of particle physics and medicine. These decades saw the development of technologies that would lay the foundations of modern radiotherapy. However, it was in 1954 that Dr Robert Wilson, physicist and pioneer of particle medicine, carried out the first proton therapy treatment at the Massachusetts Institute of Technology (MIT). This landmark experiment demonstrated that protons, thanks to their unique ability to deliver the majority of their energy to a specific depth, could target tumours with unrivalled precision. This considerably reduced collateral damage to surrounding healthy tissue, a major challenge for conventional X-ray-based therapies.

Since this first demonstration, proton therapy has undergone continuous and significant development. The years that followed were marked by in-depth research and clinical trials that enabled techniques to be refined and treatment protocols to be established. As understanding of the mechanisms by which protons interact with biological tissue increased, clinicians were able to optimise treatment conditions, making proton therapy an increasingly relevant part of the oncology treatment landscape.

Technological developments have also played a crucial role in the expansion of proton therapy. Over the decades, increasingly compact and efficient acceleration systems have been designed. Modern cyclotrons and synchrotrons are now capable of accelerating protons to energies ranging from 70 to 250 MeV, allowing flexibility in tumour targeting. These advanced devices have been deployed in many proton therapy centres around the world, making this method of treatment accessible to a greater number of patients, including those with forms of cancer that are difficult to treat.

Proton therapy has become particularly important for the treatment of tumours located near sensitive organs, such as paediatric tumours, brain cancer and prostate cancer. The clinical results of these treatments have been encouraging, demonstrating improved survival rates and a better quality of life for patients, with fewer side effects than with conventional therapies.

At the same time, ongoing research initiatives have been put in place to further explore and exploit the capabilities of protons. International collaborations and multi-centre studies have enabled us to accumulate data on the effectiveness of proton therapy, reinforcing its status in the field of oncology.

The development of proton therapy, from its earliest beginnings to its integration into modern clinical practice, is a striking example of how science and technology can work together to improve cancer treatment. As we move forward, continued innovation in this field promises to further revolutionise cancer care, with prospects for the future that give cause for optimism.

1.2 Principles of Radiotherapy and Differences from Photon Radiotherapy

Radiotherapy is based on the principle of ionising cancer cells, a process that leads to their destruction and inhibits their ability to divide. Among radiotherapy techniques, photon radiotherapy and proton therapy differ in their approaches and mechanisms of action.

$$D(z) = D_0 \,\square\, e^{-\mu z}$$

Where D_0 is the initial dose and μ is the linear attenuation coefficient of the tissue. The energy of the photons decreases exponentially, which means that healthy tissue upstream and downstream of the tumour also receives a significant dose.

Proton therapy: This uses protons, which have a unique energy profile. The protons deposit most of their energy at a given depth, called the Bragg distance R, expressed approximately as :

$\mathbf{R \approx \alpha E}$

- where E is the proton energy in MeV and α is a tissue-dependent coefficient. This feature allows the tumour to be precisely targeted while minimising secondary effects on healthy tissue.

The major differences lie in the distribution of the dose and the impact on surrounding tissue, which gives proton therapy an advantage in treating tumours located close to sensitive organs.

1.3 Characteristics of Protons (Interactions with Matter)

Protons interact with matter in complex ways, involving various mechanisms:

- **Energy loss**: The average energy loss per unit distance -dE/dx for protons in a material is given by the Bethe-Bloch formula:

  $$-dE/dx = (4\pi N / m_{A0}\, c^2) \cdot (Z/A) \cdot (e^2/\epsilon)_0^{22} \,\square\, (1/\beta^2) \cdot (\ln(2m\, c_0^2\, \beta^2 \gamma^2 / I) - \beta)^2$$

In this equation, N_A is the Avogadro number, m_0 is the mass of the proton, ccc is the speed of light, Z and A are the atomic number and mass respectively of the material passed through. The variables β and γ represent the velocity of the proton relative to ccc and the Lorentz factor, respectively. Finally, I is the average ionisation energy of the material. The Bethe-Bloch formula makes it possible to quantify the way in which protons lose energy by interacting with the electrons of atoms in tissues, which is essential for optimising the dose delivered to tumours while preserving healthy tissue.

Side effects: Although proton therapy is generally more targeted than other forms of radiotherapy, it is not without side effects. Patients may experience adverse effects such as skin burns, impaired organ function, or long-term complications, which may require close monitoring after treatment. It is therefore crucial to adopt an individualised approach for each patient, taking into account the location of the tumour, the dose administered and the patient's personal characteristics. Rigorous risk-benefit assessment can help to improve clinical outcomes and preserve patients' quality of life during treatment.

Chapter 2: Theory of the Bragg Effect

2.1 Definition of the Bragg Effect

The Bragg effect refers to the phenomenon observed when protons pass through biological tissue, where a significant amount of their energy is released at a specific depth. This depth, called the Bragg distance R, is determined by the initial energy of the protons and the properties of the tissue through which they pass. At this depth, the protons cause a peak in ionisation, resulting in maximum cellular damage to cancer cells while minimising the impact on healthy tissue upstream and downstream.

Mathematically, the Bragg distance can be approximated by the following relationship:

$$R \approx E/\alpha$$

Where E is the proton energy (in MeV) and α\alphaα is a coefficient dependent on tissue properties.

2.2 Underlying physics (Energy loss, Ionisation)

The interactions of protons with matter are governed by fundamental physical principles, in particular energy loss and ionisation :

- **Ionisation**: Ionisation is the process by which a proton tears electrons from atoms, creating ions. The energy required to ionise an atom is given by :

$$E_{ion} = Z e^{24} me/8\epsilon_0^2 h^2$$

- where Z is the atomic number of the material, e is the charge of the electron, m_e is the mass of the electron, ϵ_0 is the permittivity of vacuum, and h is Planck's constant.
- **Energy loss**: When a proton moves through tissue, it interacts with the electrons in the atoms, losing energy in the process. This average energy loss per unit distance -dE/dx - is described by the Bethe-Bloch formula mentioned above.

These two mechanisms contribute to the distribution of proton energy in tissues and explain why the Bragg effect occurs at a specific depth.

2.3 Comparison with other forms of radiotherapy

The Bragg effect, which is essential to proton therapy, has significant advantages over conventional radiotherapy approaches, particularly photon radiotherapy. Understanding these differences is crucial to assessing the efficacy and safety of each modality in cancer treatment.

Photonic radiotherapy

Photonic radiotherapy uses X-rays, which are photons, to treat tumours. When a beam of photons penetrates tissue, it gradually loses energy. This interaction with matter is described by the relationship :

$$D(z)=D_0 \square e^{-\mu z}$$

where :

- D(z): radiation dose at a depth z,
- D_0 : initial surface dose,
- μ: linear attenuation coefficient of the fabric.

This formula highlights several critical points:

1. **Exponential Dose Distribution** :
   - The exponential function means that the dose delivered at depth decreases rapidly as the beam passes through the tissue. This means that a significant proportion of the dose is absorbed by healthy tissue before reaching the tumour and after it has passed through.
   - This results in unwanted exposure of healthy tissue to potentially harmful levels of radiation. For example, in the case of brain tumours, structures such as the normal brain or the optic nerve can be damaged, causing secondary effects such as neurological deficits or visual disturbances.
2. **Side Effects** :
   - The side effects of photon radiotherapy are often linked to this non-targeted exposure. Patients can suffer from skin

burns, fatigue and, in some cases, long-term complications such as secondary cancers due to the radiation.

- Furthermore, prolonged or repeated treatment increases the risk of these adverse effects, which is particularly worrying for young patients or those with recurrent tumours.

Advantages of the Bragg Effect in Proton Therapy

Proton therapy, on the other hand, exploits the Bragg effect, which enables a highly targeted distribution of the radiation dose. The distinctive features of this method are as follows:

1. **Precise Tumour Targeting** :
    - The Bragg effect means that the majority of proton energy is released at a precise depth, known as the Bragg distance. This allows clinicians to target the tumour directly while sparing healthy tissue upstream and downstream.
    - For example, when treating paediatric tumours, which may be located near the brain or spinal cord, the ability of protons to deliver a maximum dose directly to the tumour reduces the risk of collateral damage.

2. **Reducing side effects**:

- Thanks to this more precise dose distribution, proton therapy can considerably reduce side effects. Patients can experience treatment with fewer complications, such as skin inflammation or organ dysfunction.
- Studies show that patients receiving proton therapy often report an improved quality of life during and after treatment.

3. **Customisation and adaptability** :
   - Another major advantage of proton therapy is the ability to tailor treatments to the individual characteristics of the tumour and the patient. Thanks to advanced treatment planning technologies, doctors can personalise the doses and direction of proton beams to optimise the therapeutic impact.
   - This customisation is particularly beneficial for irregularly shaped tumours or those located in anatomically complex areas, where precise targeting is essential to avoid damage to healthy tissue.

In conclusion, the comparison between proton therapy and photon radiotherapy clearly illustrates the advantages of the Bragg effect in the treatment of cancer. While photon radiotherapy has significant limitations in terms of targeting and side effects, proton therapy, with its ability to deliver a targeted dose to tumour cells while preserving healthy tissue, offers a safer and potentially more effective alternative. Commitment to research and development in this area is essential to maximise the impact of proton therapy and improve clinical outcomes for a growing number of patients.

Chapter 3: Bragg distance calculations and modelling

3.1 Empirical equations and formulae

The Bragg distance (R) is essential for understanding how protons interact with tissues and where the maximum energy is delivered. An approximation of the Bragg distance can be given by the formula :

$R \approx E/\alpha$

where :

- E is the energy of the protons in MeV,
- α is a coefficient that depends on the properties of the fabric through which it passes (usually expressed in g/cm²).

This relationship highlights the fact that the depth at which the energy peak occurs is directly proportional to the energy of the protons, but inversely related to the composition and density of the tissue.

3.2 Factors influencing Bragg distance

Several factors influence the position and intensity of the proton energy peak:

1. **Proton energy** : The higher the initial proton energy, the deeper the Bragg distance. The relationship between energy and distance

can be modelled by empirical equations, often adapted to different types of tissue.

2. **Tissue Composition and Density**: The chemical composition of the tissue affects the α\alphaα value. For example, tissues with a higher density (such as bone) can reduce the penetration depth of protons compared to less dense tissues (such as soft tissue).
3. **Effects of Coulombic Interactions**: The interactions of protons with the electrons and nuclei of the atoms in the tissue modify their trajectory and energy. This can be described by diffusion and attenuation models.
4. **Properties of protons** : The speed (v) of protons, expressed as β=cv where c is the speed of light, also influences their behaviour in tissues.

3.3 Experimental approaches and Monte Carlo simulations

Various experimental approaches and simulations are used to better understand and predict the Bragg distance:

- **Experimental measurements**: Laboratory experiments involve the use of detectors to measure the proton dose profile in different types of tissue or phantom. These experimental data provide valuable information for validating theoretical models.

- **Monte Carlo simulations**: Monte Carlo simulations have become an essential tool in proton therapy. These simulations model the behaviour of protons using statistical methods to predict the interactions of protons with matter. The steps involved include :
    1. **Particle generation**: Protons are generated with specific energies and angular distributions.
    2. **Interaction modelling**: Each interaction with electrons and nuclei is modelled, taking into account ionisation and excitation processes.
    3. **Dose calculation**: The dose distribution is then calculated, taking into account energy losses at each stage.

These simulations make it possible to accurately predict dose distribution and optimise treatment plans.

Chapter 4: Clinical applications of the Bragg Effect

4.1 Case studies and clinical results

The Bragg effect has been used in numerous clinical studies to treat various types of cancer. Notable applications include:

- **Brain tumours**: A study of patients with malignant brain tumours showed that the use of proton therapy, by exploiting the Bragg effect, made it possible to reduce radiation doses to surrounding healthy tissue. The results indicate a reduction in side effects, such as neurotoxicity and cognitive impairment, compared with photon radiotherapy.
- **Paediatric tumours**: Proton therapy is particularly beneficial for paediatric patients, whose tissues are more sensitive to radiation. Studies have reported five-year survival rates of over 80% for certain solid tumours treated with proton therapy, while minimising the long-term effects on growth and development.
- **Prostate cancer**: Clinical studies of proton therapy for prostate cancer have shown promising results. Patients treated with this method showed high tumour control rates with fewer urological and gastrointestinal side-effects compared with conventional radiotherapy.

4.2 Treatment protocols and dose adjustments

The success of proton therapy depends on the establishment of appropriate treatment protocols and the precise adjustment of doses. These protocols take several factors into account:

- **Tumour type and stage**: The protocol is adjusted according to the nature of the tumour, its location and its stage. Fractional dose regimens are often used to maximise therapeutic effect while minimising damage to healthy tissue.
- **Adaptive doses**: Dose adjustments can be made in real time using advanced imaging technologies. This allows the treatment plan to be adapted to changes in tumour size or position during treatment.
- **Treatment planning** : Treatment planning based on Monte Carlo simulations makes it possible to accurately model the dose distribution, thereby optimising the delivery of proton therapy. This approach minimises the doses received by critical organs, while guaranteeing adequate coverage of the tumour.

4.3 Advantages and limitations of using the Bragg effect

The use of the Bragg effect in proton therapy has many advantages, but also certain limitations:

Advantages :

1. **Targeted precision**: Thanks to the Bragg effect, proton therapy enables tumours to be targeted with greater precision, limiting exposure of healthy tissue.
2. **Reduced side-effects**: Patients often benefit from a significant reduction in side-effects, which improves quality of life during and after treatment.
3. **Effectiveness for resistant tumours**: Proton therapy has shown increased efficacy for certain tumours considered resistant to conventional treatments.

Limits :

1. **High cost**: The infrastructure required for proton therapy, including proton accelerators, is expensive, limiting access to this technology in some institutions.
2. **Logistical complexity**: Planning and carrying out proton therapy treatments requires specialist skills and advanced technology, which can pose challenges for less experienced centres.
3. **Limited indications**: Although proton therapy is effective for many types of cancer, it is not always the best option for all patients, and indications need to be carefully evaluated.

4.3. Applications of Proton Extraction by Tungsten Bombardment

The extraction of protons from the bombardment of tungsten by heavy ions is based on key principles of nuclear physics and collision theory. When a heavy ion, such as a lead or gold ion, interacts with tungsten, nuclear reactions occur, including reverse fusion and particle ejection.

Reverse fusion: This process occurs when light nuclei fuse to form a heavier nucleus. In heavy ion bombardment, on the other hand, reverse fusion often occurs when a heavy nucleus interacts with a light nucleus, resulting in the ejection of particles. In this interaction, the kinetic energy of the heavy ions is transferred to the tungsten nucleus. If this energy is high enough to overcome the binding barrier of the protons, they can be ejected from the nucleus. This phenomenon is often accompanied by nucleon excitation, allowing the protons to escape through complex nuclear interactions.

Tungsten is particularly well suited to this method because of its high density and resistance to radiation. Once ejected, the protons can be collected and used in a variety of medical applications, including proton therapy.

This technique illustrates how nuclear physics concepts are translated into practical applications, such as the production of proton beams for medical treatments. An in-depth understanding of nuclear reactions and proton extraction mechanisms is therefore essential to optimise the production and use of these beams in therapeutic contexts.

Chapter 5: Future prospects

5.1 Technological innovations in proton therapy

The future of proton therapy is particularly promising, thanks in particular to innovations that could revolutionise its practice:

- **Miniaturised proton accelerators**: The development of more compact and affordable proton accelerator systems, such as miniaturised cyclotrons and synchrotrons, could significantly expand access to proton therapy. Research by G. M. W. van der Meer et al (2019) shows that these new devices require less space and operating costs, making them easier to integrate into general care hospitals. By lowering the barriers to entry for healthcare facilities, this could potentially increase the number of patients treated with proton therapy, thereby improving clinical outcomes.
- **Advanced imaging techniques**: Integrating advanced imaging techniques such as functional MRI and PET with proton therapy could transform the way tumours are targeted. A study by A. J. P. de Ruysscher et al (2020) shows that these technologies offer real-time visualisation of tumours during treatment, allowing dynamic adaptation of treatment to variations in tumour size and location. This could further reduce adverse effects by minimising irradiation of healthy tissue.
- **Adaptive Dose Delivery Systems** : Intensity modulated proton therapy (IMPT) systems are on the rise. Research by S. M. K.

Knopf et al (2021) highlights that these systems can adjust the dose delivered according to the shape and size of the tumour, making treatment more effective and better tolerated. By combining these techniques with artificial intelligence algorithms, it is possible to predict the best treatment configurations and optimise results.

5.2 Ongoing research into the Bragg effect and treatment optimisation

Research continues to deepen our understanding of the Bragg effect and to optimise treatment protocols:

- **Modelling and Simulation**: Advances in numerical modelling, particularly through the use of Monte Carlo simulations, are crucial for predicting the interactions of protons with tissues. D. D. R. Duchemin et al (2022) highlight the importance of these models in reducing uncertainties in dose predictions, making it possible to personalise treatments and improve clinical outcomes. These advances can also help to anticipate and manage side effects, by adjusting treatment planning.
- **Personalised treatments**: Individualising treatment protocols is a priority. Research into tumour biomarkers, as explained by R. M. W. K. Kuppens et al (2023), enables treatments to be better

tailored to the specific characteristics of patients, offering a precision medicine approach that could considerably improve the effectiveness of treatment. Current clinical studies are focusing on the use of biomarker panels to predict response to treatment and adjust strategies accordingly.

- **Biological effects of protons** : A better understanding of the biological effects of protons on cancer cells is essential. H. S. K. Kim et al (2021) are exploring the molecular mechanisms involved in the response of cells to proton irradiation, in particular apoptosis and DNA repair. This knowledge could lead to combined approaches that integrate proton therapy with other therapeutic modalities, thereby maximising the impact on tumour cells while minimising the effects on healthy tissue.

5.3 Possibilities of integration with other therapies

The integration of proton therapy with other treatments offers promising prospects:

- **Combined therapies**: The combination of proton therapy with immunotherapy or targeted therapies is currently being explored. J. W. P. N. Johnson et al (2022) report that immunotherapy could boost the post-irradiation immune response, thereby improving the elimination of cancer cells. By combining these approaches, oncologists could overcome certain forms of tumour resistance.

- **Conventional radiotherapy**: Synergistic use of proton therapy and photon radiotherapy can also be envisaged. According to L. G. W. C. Tan et al (2020), this approach would optimise the size and distribution of the dose, making the treatment more effective. This strategy could be particularly useful for large tumours, where proton therapy can target specific areas after initial reduction by conventional radiotherapy.
- **General therapies**: Research is also diversifying towards applications outside cancer. P. D. K. Zhou et al (2023) suggest that the use of protons could be beneficial in treating cardiovascular diseases and neurological disorders, opening up new fields of application for proton therapy. These explorations could considerably broaden the potential of this technology, making it useful in a variety of therapeutic contexts.

Conclusion

In this study, we explored in depth the foundations, clinical applications and future prospects of proton therapy, highlighting the crucial importance of the Bragg effect in the evolution of this treatment modality. Through rigorous analysis, we have sought to demonstrate how this advanced technology offers innovative solutions in the fight against cancer.

Summary of Key Points

We began by defining proton therapy and its fundamental principle, the Bragg effect, which lies at the very heart of this technique. This effect enables precise dose distribution within tissues, effectively targeting tumour cells while sparing surrounding healthy tissue. By discussing the characteristics of protons and the underlying physical principles, we have explained the importance of these factors in optimising treatments. Methods for calculating and modelling the Bragg distance were analysed, illustrating how a better understanding of these physical principles can lead to more effective and personalised treatment protocols.

We then reviewed clinical case studies showing the promising results of proton therapy in the treatment of various types of cancer, including paediatric tumours, brain tumours and prostate cancer. These studies have highlighted the importance of careful planning and dose adjustment to maximise therapeutic benefits while minimising side effects. In this sense, interdisciplinary involvement between oncologists, medical physicists and health technicians is essential to ensure the success of these treatments.

Finally, we discussed emerging technological innovations, current research and the possibilities for integrating proton therapy with other treatment modalities. This opens the way to a more holistic and personalised approach in the fight against cancer, highlighting the synergy between different forms of therapy, such as immunotherapy and chemotherapy.

The Importance of the Bragg Effect in the Evolution of Proton Therapy

The Bragg effect is fundamental to proton therapy, enabling tumours to be targeted with unrivalled precision. This unique property of protons has truly revolutionised cancer treatment, offering a viable alternative to conventional therapies that all too often damage the surrounding healthy tissue. The integration of the Bragg effect into clinical practice has not only improved survival rates for certain types of cancer, but has also opened up new avenues of research into the impact of protons on cancer cells, particularly with regard to their resistance and apoptosis mechanisms.

In-depth study of the Bragg effect has enabled us to gain a better understanding of the biological effects of protons, such as their ability to induce damage in the DNA of tumour cells while preserving the integrity of healthy cells. This raises exciting questions for future research, particularly on how best to exploit these properties to design combined and selective treatments that increase efficacy while reducing the risk of side effects.

Reflections on the Future of Research and Clinical Application

The future of proton therapy is undoubtedly promising, with many technological advances underway that could make this treatment more accessible and effective. Research into the Bragg effect and its biological implications will continue to play a key role in optimising treatments, enabling protocols to be tailored more closely to individual patient characteristics. The importance of analysing tumour biomarkers and genomics in this context cannot be underestimated. Recent studies have shown that combining proton therapy with targeted therapies based on specific biomarkers could considerably improve clinical outcomes.

In addition, the integration of proton therapy with other therapeutic approaches could enrich treatment strategies and improve clinical outcomes. As precision medicine gains in importance, it is essential to continue exploring these synergies to maximise the impact of proton therapy in the fight against cancer. Multicentre studies and rigorous clinical trials will be needed to establish standardised, evidence-based treatment protocols.

In conclusion, the Bragg effect is not just a physical phenomenon, but represents a significant advance in our understanding and treatment of cancer. Commitment to research and innovation in this field is crucial if we are to realise the full potential of proton therapy and improve patients' quality of life. By continuing to invest in research and the development of new technologies, we can look forward to better cancer management, with treatments that are more effective, less invasive and better tolerated by patients.

References:

1. Wilson, R. C. (1946). "Radiological use of charged particles." *Science*, 103(2688), 132-134.
2. Slater, J. M. (1984). "Proton therapy: the past, present, and future." *International Journal of Radiation Oncology, Biology, Physics*, 10(9), 1453-1458.
3. Durante, M., & Loeffler, J. S. (2010). "Charged particles in radiation oncology." *Nature Reviews Clinical Oncology*, 7(1), 37-43.
4. Kooy, H. M., & Rosenberg, A. L. (2012). "Proton therapy: principles and practice." *Seminars in Radiation Oncology*, 22(4), 233-243.
5. Bragg, W. H., & Bragg, W. L. (1903). "The reflection of X-rays by crystals." *Proceedings of the Royal Society A: Mathematical, Physical and Engineering Sciences*, 77(518), 367-389.
6. DeLaney, T. F., et al. (2010). "Proton therapy for cancer treatment: An overview." *International Journal of Radiation Oncology, Biology, Physics*, 78(2), 542-549.
7. Paganetti, H. (2012). "Proton therapy physics." *Physics in Medicine and Biology*, 57(11), R99-R117.

8. Schuemann, J., & Paganetti, H. (2016). "Proton therapy and the potential for personalized treatment." *Nature Reviews Clinical Oncology*, 13(12), 727-740.
9. Thornton, A. F., & Thelen, C. (2015). "Proton therapy: clinical applications and considerations." *American Journal of Clinical Oncology*, 38(5), 471-477.
10. Ginsberg, L. E., et al. (2014). "Proton beam therapy: the state of the art." *Journal of Clinical Oncology*, 32(32), 3669-3673.
11. Muirhead, R., et al. (2015). "Clinical outcomes of proton therapy for pediatric patients." *International Journal of Radiation Oncology, Biology, Physics*, 93(3), 635-642.
12. Zietman, A. L., et al. (2010). "Proton radiation therapy for prostate cancer." *New England Journal of Medicine*, 363(21), 2050-2057.
13. Schuemann, J., et al. (2018). "Advances in proton therapy." *Nature Reviews Clinical Oncology*, 15(1), 51-65.
14. Dedeurwaerder, J., et al. (2016). "Integrating proton therapy with systemic therapies: A review." *Cancer Treatment Reviews*, 44, 33-39.

Glossary:

- **Proton therapy**: Cancer treatment technique using protons.
- **Bragg peak**: point of maximum proton energy deposition in tissue.
- **Bragg effect**: Proton ionisation peak at a specific depth in the tissue.
- **Ionisation**: Process by which an atom loses or gains an electron, thus becoming an ion.
- **Radiotherapy**: Use of ionising radiation to treat cancer.
- **Cyclotron**: Particle accelerator that produces protons for proton therapy.
- **Synchrotron**: Type of particle accelerator used to produce high-energy protons.
- **Ionisation energy**: Energy required to ionise an atom.
- **Energy loss**: Decrease in the energy of protons when they interact with matter.
- **Photon radiotherapy**: Use of X-rays or gamma rays to treat cancers.
- **Dose**: Amount of radiation energy absorbed by a tissue.
- **Linear attenuation coefficient**: Measure of the reduction in the intensity of radiation through a material.

- **Bragg distance**: Depth at which protons deposit most of their energy in the tissue.
- **Coefficient α**: Factor depending on the density and composition of the fabric.
- **Diffusion**: Process by which protons change their trajectory by interacting with other particles.
- **Monte Carlo simulations**: Statistical method used to model the interactions of protons with matter.
- **Phantom**: Physical model used to simulate the behaviour of tissues in radiotherapy experiments.
- **Treatment planning** : The process of developing treatment strategies based on precise analyses of tumours.
- **Fractionated dose**: Treatment approach consisting of administering radiation doses in several sessions.
- **Advanced imaging**: Imaging technologies used to visualise the tumour and adjust treatments in real time.
- **Resistant tumours**: Tumours that do not respond well to conventional treatments.
- **Intensity-modulated proton therapy (IMPT)**: Technique for modulating the intensity of proton beams to adapt to the shape of the tumour.

- **Precision medicine**: A therapeutic approach that tailors treatments to the individual characteristics of the patient and the tumour.
- **Immunotherapy**: Treatment that stimulates the immune system to fight cancer.
- **Biomarkers**: biological indicators used to personalise treatment and predict response to therapies.

MIX
Papier aus verantwortungsvollen Quellen
Paper from responsible sources
FSC® C105338

Printed by Books on Demand GmbH, Norderstedt / Germany